Animals That Glow

ANIMALS THAT GLOW

by
Judith Janda Presnall

Franklin Watts
New York / Chicago / London / Sydney
A First Book

Dedication
In loving memory of my dad
Edward G. Janda
who took pride in my accomplishments

Acknowledgments
Thanks to my family and writing colleagues for
their patience, support, and encouragement.

Cover photograph courtesy of James E. Lloyd.

Photographs courtesy of: Nancy Sefton/Photo Researchers, 8; Jean-Marie Bassot, Jacana/Photo Researchers, 11; SatoshiKuribayashi/Nature Production, 13; James E. Lloyd, 16; Mitsuhiko Inamori/Nature Production, 18; © Robert F. Sisson, All Rights Reserved, 20, 23; Brian Brake/Photo Researchers, 25; Kjell B. Sandved/Photo Researchers, 27; Rowland M. Shelley, North Carolina State Museum of Natural Sciences, 32; Norbert Wu, OSF/Animals Animals, 35; Fred McConnaughey/Photo Researchers, 38; Clay H. Wiseman/Animals Animals, 41; James G. Morin, University of California, Los Angeles, 45; John Lidington/Photo Researchers, 48; Dr. Paul A. Zahl/Photo Researchers, 50; Dick Rowan, Photo Researchers, 52; George Whiteley/Photo Researchers, 54.

Library of Congress Cataloging-in-Publication Data
Presnall, Judith Janda.
Animals that glow / by Judith Janda Presnall.
  p. cm.—(First books)
Includes bibliographical references (p.) and index.
Summary: A study of insects and other animals that are
bioluminescent, including fireflies, glowworms, and squids.
ISBN 0-531-20071-X (hc : lib. bdg.)   QH641.P74 1993
ISBN 0-531-15672-9 (paperback)   591.19'125—dc20
1. Bioluminescence—Juvenile literature.   92-25529 CIP
[1. Bioluminescence.]   AC
I. Title. II. Series.

OCT 2012

Contents

Animals That Glow: Lights from Nature

Imagine a worm that resembles a toy train zooming through the night with its headlight and windows aglow. Think about flying beetles with magic lanterns that blink codes and create a fairyland of lights. Envision glowworms that fill a cave with tiny blue lights. Visualize a fish with built-in flashlights.

Nature makes all of this possible through **bioluminescence**, which is light that comes from living organisms. **Bio** means "living," and **luminescence** means "giving off light."

Animals that glow can live in caves, grass-lands, or forests; warm shallow ocean waters; or the deep, dark sea — or even your own yard and garden. There are thousands of bioluminescent **species**, or kinds, of animals. You will be reading about some of these creatures that have this remarkable ability to create light.

A jellyfish shows off its lacelike cap.

Some luminescent animals give off light all of the time. Other luminescent animals flash their light whenever they choose. The lights are used for different purposes. They may be used to communicate, to **camouflage**, to guide, to attract mates, to lure **prey**, or to frighten enemies.

Bioluminescent light differs from electric light and light from fire and the sun, which are warm. Bioluminescent light is a cold light. It comes from two special substances, **luciferase** and **luciferin**, in the bodies of luminescent animals. These substances mix with **oxygen** and another chemical, **ATP**, to produce light. ATP stands for adenosine triphosphate, which is a chemical energy found in all living things.

Bioluminescent animals live both on land and in the sea. On land, luminescence is relatively rare. In fact, there are no luminescent birds, mammals, or reptiles. The most common land animals that glow are fireflies, **glowworms**, and millipedes. Most glowing land animals live in cool, damp, and dark places.

More than 80 percent of all ocean animals show some bioluminescence. In fact, under the pitch black depths of 6,000 feet (1.8 km), every known swimming creature is luminescent in one way or another.

Some fish may not have the special light-producing chemicals themselves. Instead, the **bacteria** living inside or outside their bodies have these chemicals. **Luminous** fish have a variety of lights located along their sides, on their heads, in rows on their bellies, or all over their bodies. Lights of pink, red, white, purple, and blue, displayed by thousands of species of fish, make the underwater world a breathtaking parade of color.

Scientists believe that only saltwater organisms can produce light. The vast list of marine bioluminescent creatures includes species of fish, jellyfish, worms, snails, clams, squid, octopuses, coral, bacteria, and dinoflagellates.

Many different species exist within a family. Although the species belong to the same family, they may not have all of the same characteristics, just like members of your own family. For example, all fireflies are not identical. Each species has its own flashing code. All squid do not luminesce in the same way. Some have the lights inside their bodies, others release luminescent chemicals into the water, and still others have luminous bacteria growing on their skin. Glowing animals, however, do have one thing in common — the presence of luciferase and luciferin.

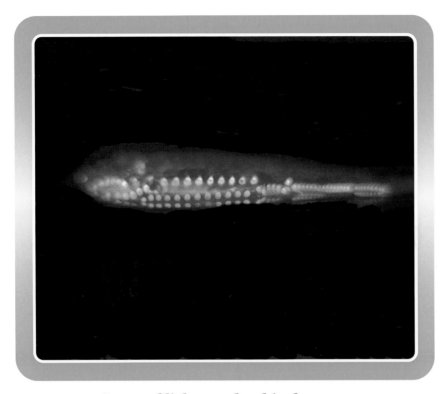

Rows of lights make this deep sea bristlemouth fish easy to spot in the pitch dark ocean.

Luminescence is almost always in lower life forms. Fish are the most developed. Moving downward in complexity are insects, millipedes, **crustaceans**, worms, squid, **fungi**, bacteria, and finally, dinoflagellates.

CHAPTER 2

Fireflies:
Code-Blinking Beetles

During hot summer evenings you may see flickering yellow or green lights in your garden or on your lawn. These fairylike insects are fireflies. But they are not really flies. They are beetles with light organs under their tails.

The fireflies, or **lampyrids**, that you see blinking and flashing in the air are mostly males. They are looking for a mate. Each species has its own code. The male blinks his half of the code. The female, usually perched on a blade of grass, answers with the other half. Flight and flash patterns vary with each species. Some bob and shimmer, others float upward weaving "J" or "S" forms of light, and still others change the brightness of their flash.

Firefly courtship starts when the male flies through the air flashing a signal from his tail. The female answers the male by twisting and

*Flashing its tail light, this firefly
signals for a mate.*

turning her tail and flashing back. She shines
her light in all directions, similar to a searchlight
at an airport. Her responding flash must be the
proper pattern for her species and match the cor-
rect timing of flashes. This tells the male she is
of the same species. Before the male flies to the
female and they mate, these flashing signals may

be repeated five to ten times. Sometimes it takes several hours to find a mate.

The light flash of the firefly occurs when chemicals in the tail, called luciferase and luciferin, combine with oxygen from the air.

There are more than two thousand named firefly species throughout the world. Fireflies are found on every continent except Antarctica. The most common U.S. fireflies, *Photinus* and *Photurus*, are usually found in the Midwest and the East.

Florida, with its warm climate, has fireflies most months of the year. The temperature also affects the speed of flashes. For instance, at 85° Fahrenheit (29° Celsius) flashing is about twice as fast as it is at 70° F (21°C).

The firefly life **cycle** goes through egg, **larva**, pupa, and adult stages. The female firefly lays from forty to one hundred soft, round eggs in loose soil or in a burrow. About three weeks later, the eggs hatch into tiny brown worms, called glowworms or luminous larvae.

When the glowworm hatches, it is about the size of a pencil point (1 mm), but already this tiny worm has feelers, eyes, strong curved jaws, six legs, and two spots of light on its tail. The baby glowworm lives in damp places underground or beneath logs, bark, and fallen leaves.

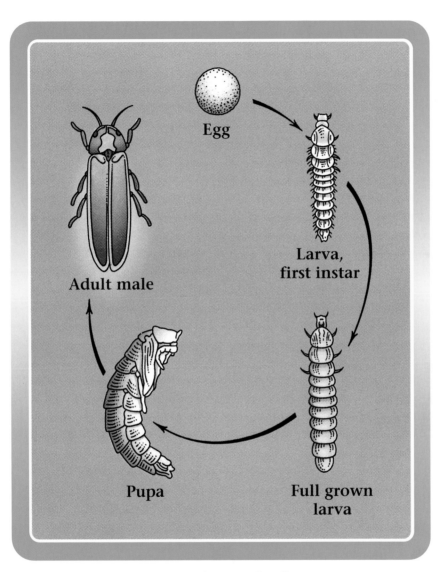

Egg

Larva,
first instar

Adult male

Pupa

Full grown
larva

Life cycle of a firefly

A glowing larviform female protects her cluster of eggs.

The glowworm sleeps during the day. It spends the evening searching for food to satisfy its huge appetite. Snails, slugs, and earthworms are its favorite meals. In farming areas and gardens where snails and slugs are both pests and harmful to crops, glowworms are welcome dwellers.

You are probably surprised that the glowworm can eat food larger than itself. It uses its strong jaws, called **mandibles**, to bite its prey several times. Poison flows through its fanglike pincers, making the prey sleepy and helpless. The prey's muscle fibers change to a thick liquid, which the glowworm eagerly sucks up.

This larval or glowworm stage is the longest in the firefly's life cycle. Depending on the species and location, the entire cycle can range from three months to three years.

The light of the glowworm is quite soft. If you put a glowworm on a page of a book in a dark room, you could probably read only one word.

When the air grows cold, a glowworm finds a home under a stone, in a log, or in a hollowed-out space in the soil. When the glowworm reaches its full growth, it is ready to enter the pupal stage. It builds an igloo-shaped chamber of mud, usually at ground level or below the ground, and there the pupal stage begins. The glowworm's body puffs up. About two weeks later, the larva changes into a firefly.

Adult fireflies live only from one to twenty-one days. During this time, the firefly's chief job is to mate with another firefly. Once the flashing ritual and mating are over, a new life cycle begins with the female laying her eggs.

An unusually spectacular event has been seen in Thailand, Burma, Malaysia, and the

Branches and leaves shimmer with light from millions of fireflies.

Philippines. Millions of male fireflies, of the group scientists call *Pteroptyx*, gather in trees growing in water. Flickering 100 to 120 synchronized flashes per minute, they put on a shimmering show of light. These glittering "Christmas trees" of male fireflies attract females and become huge **breeding** areas.

CHAPTER 3
The Railroad Worm: Nature's Toy Train

Few people have seen the railroad worm, but when they do, it is a sight to see. The railroad worm looks like a toy train when it lights up, hence the name. When the worm is touched its head glows a fiery red, like a headlight. And along each side of its body are eleven spots of shining, greenish yellow lights that look like the windows on a train. The railroad worm is the larva of the beetle *Phrixothrix*.

These larvae live in the grasslands of Central and South America. Most are found in Paraguay and Brazil. They spend much of their lives hidden in the earth or under logs and rocks. Their reddish brown bodies are difficult to see on soil.

The female larva can be up to 2 inches (5 cm) long, while the male is only half that size. The railroad worm turns on its lights only when

The bright red head and eleven glowing yellow segments of the rare railroad worm are a special sight.

it is aroused by actual contact with some other creature. It can control its side lights separately, so much so that while the head is lit up, individual "window" lights may blink.

After hatching from eggs, railroad worms live a year or longer in the larva form. The pupal stage lasts about twenty-five days for the male and about ten days for the female.

After casting off their pupal skins, they emerge as mature adults. The male adult is a small winged beetle with tiny hairs on its body. In fact, *Phrixothrix* means "with bristling hair." The male emits a greenish yellow light from his abdomen, but his head no longer glows red.

The adult female looks almost the same as she did in the larval stage. Only now she has reproductive organs. She is again capable of glowing both red and yellow. Spending most of her brief adulthood in an underground cell, she comes to the surface only to mate.

A faint scent rather than luminous signals attracts males to females. Both insects glow brightly when they mate. Soon after, the female lays about three dozen pearly white eggs, which soon turn reddish brown.

Because they live underground, adult females are difficult to find. But the male beetles are quite easy to catch because they live above ground and are attracted to light.

The behavior of railroad worms changes in captivity. When placed in a container, they light up readily if someone taps on their jar. Noises or vibrations, such as a closing door, also cause them to light up. If two or more are kept in the same jar, they attack and may even eat each other. Railroad worms also give off a brilliant light when they attack millipedes offered as food. They curl around the millipede's body and bite it. As their sharp jaws sink into their victim, they secrete or regurgitate (throw up) a dark fluid, possibly a poison or a stomach **enzyme**, from the mouth.

Scientists believe that railroad worms produce light in the same way that their relatives, the fireflies, do. Their bodies combine oxygen with the chemicals luciferase and luciferin, and that causes the glow.

The red light of the railroad worm is unique among insects. Scientists believe the color is determined by the makeup of different enzymes in the cells of the light organs. In the railroad worm, some enzymes cause the body to produce yellow lights, and other enzymes cause the head to produce red light.

*A nonluminescing railroad worm
with her eggs*

Glowworms:
Heavenly Stars in Caves

A famous tourist attraction in New Zealand is a cave called the Glowworm Grotto, located in the Waitomo Caves. Entering the cave on a small boat, tourists are treated to thousands of starlike, greenish blue lights. The lights come from larvae, or glowworms, that live on the ceiling of the cave.

These glowworms differ from those of the firefly or beetles in other parts of the world. The New Zealand glowworm, *Arachnocampa luminosa*, is the larval stage of a small fly or gnat.

Living conditions for this fly include damp air, a windless location, darkness, a suitable hanging surface, and a supply of flying insects for food.

The life cycle of this fly lasts about eleven months. Throughout the year, all four life stages

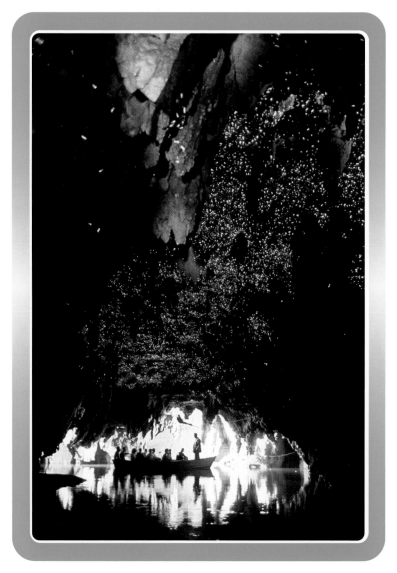

*Inside a Waitomo Cave, a boat of tourists
admires a glittering ceiling.*

are present in the cave. The eggs are sticky when laid and cling to the ceiling of the cave. There is no luminescence in the egg. During the next three stages, however, luminosity is present at different times.

After three weeks, the glowworm hatches. It immediately begins to build a nest. The nest is a clear hollow tube that hangs from the ceiling by threads. It is made of **mucus** (saliva) from glands in the larva's mouth. The tubular nest is about 5 inches (13 cm) long. It can have as many as seventy sticky threads hanging from the nest, each with drops of mucus that resemble pearl beads. The sticky threads are 6 to 20 inches (15 to 50 cm) long and are used to capture flying insects to eat. Both the nest and the glowworm are transparent; you can even see the glowworm's insides.

The larval stage is the only time the glowworm eats. That is why this stage is much longer than the others. It has to store up enough food to feed the pupa and adult, and if a female, her eggs.

The glowworm will grow to about 1½ inches (4 cm) during the next nine months. Development to the pupa stage, however,

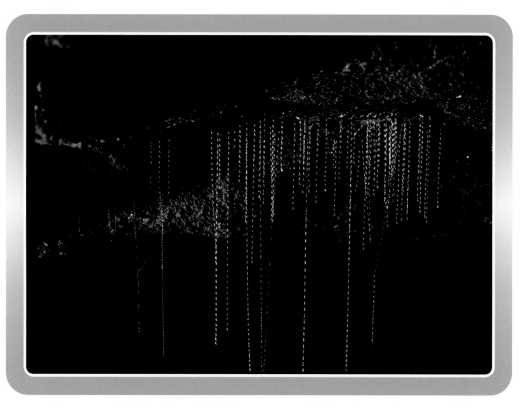

Sticky beaded lines snare flying insects
for hungry larvae.

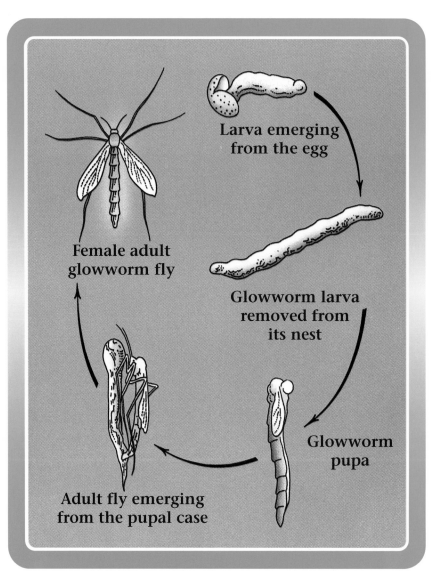

Larva emerging from the egg

Glowworm larva removed from its nest

Glowworm pupa

Adult fly emerging from the pupal case

Female adult glowworm fly

Lifecycle of the New Zealand Fly

depends strictly on body weight or size, and this is related to the availability of food. A glowworm could be a larva for nine months, or it could live as long as eighteen months before growing into the pupa stage if it couldn't find enough food. It can survive several months without any food.

The glowworm uses its light to attract food. The light organ is located on the tail of the larva. As in the firefly, its bioluminescence is caused by the mixture of luciferase, luciferin, and oxygen.

The glowworm eats any unfortunate creature that gets caught in its hanging threads, mostly tiny flies and mosquitoes. By means of special sensory organs, the larva feels vibrations and knows when food is caught in its line. Keeping its tail in the nest for stability, the larva slides down its sticky thread toward its food. Using body contractions, the larva lifts the stringed droplets on the line until it reaches the prey. It is believed that saliva acid on the lines paralyzes the captured prey. The larva bites and kills the paralyzed insect. Then it may eat the whole insect or suck out its **innards**.

When the larva has grown large enough, it is ready to become a pupa. The larva shrinks to the size of a large grain of rice and hangs on a long thread from its nest. During the next twelve days in its pupal case, it changes into an adult fly.

The fragile, long-legged fly emerges head-first and looks like a daddy longlegs spider. The adult has two purposes: (1) to reproduce and (2) to scatter the species, since the flies are the only stage that can travel very far.

After the female mates, she flies and lays her eggs on the cave ceiling. She will lay about 130 eggs in clumps of 40 to 50. It can take up to twenty-four hours to lay her eggs because each egg is laid separately. The female usually dies after laying her eggs. The flies live only one to four days.

Visitors to the narrow 100-foot (30-m) cave must be silent. Any sudden noise, cough, sneeze, or even a camera flash will cause the larvae to turn off their lights instantly, plunging the entire cave into darkness. After a few moments of silence, the tiny lights slowly blink on again.

CHAPTER 5

Millipedes: Warning Glow

Millipedes have long rounded bodies with many segments and legs. Among the eight thousand millipede species, only a few are luminous. One such species, the *Luminodesmus sequoiae*, lives in the Sequoia forests of California. Although some species have eyes, this one is blind. Its entire body shines greenish white continuously after it hatches.

The *Luminodesmus sequoiae's* glowing light is probably an advance warning sign to the millipede's enemies that it can hurt or kill. The millipede's defense against **predators** is a row of poisonous glands down each side of its body. The secreted yellow-brown liquid is very foul-smelling. Some larger tropical millipedes actually spray their predators and can temporarily blind a chicken who wants it for a meal. Some

Millipedes protect themselves by excreting a poisonous liquid.

poisons that are discharged onto human skin will make it blacken and peel.

Millipedes eat vegetable matter. Primarily night animals, they live in moist soil, crevices, and under moldy leaves. When attacked, they often will curl into a ball. After they mate, the female lays ten to three hundred eggs (depending on species) and covers them with waste and soil.

Millipedes generally differ from centipedes in that they have two pairs of legs on each body segment. When born, they don't have all of their legs. They add on legs and segments as they grow and molt. Despite their name, millipedes don't really have a thousand legs. The most legs found on a millipede is 240.

Other light givers that creep or crawl on land include certain species of centipedes, earthworms, and snails.

Bioluminescence on land is also visible on fungi such as toadstools or mushrooms. Rotting logs or piles of leaves will also glow from bacteria and mold organisms, which are so tiny they can only be seen with a microscope. They are $1/20,000$th of an inch in diameter. It would take fifty trillion of such bacteria to generate one candlepower of light.

CHAPTER 6

Anglerfish: Gleaming Lure

Various forms of animals that live in the ocean also are bioluminescent. The chemicals luciferase and luciferin are present on either the inside or the outside, or in bacterial parasites living within the sea animals' bodies. Some marine museums have darkened tanks with displays of these fascinating animals for you to view.

Deep-sea anglerfish live in the Atlantic, Pacific, and Indian oceans at a depth of more than 1 mile (1.6 km) from the surface. Because sunlight cannot filter down that far, the water at that depth is dark as night.

Some anglerfish have a **filament**, like a fishing rod, attached to their heads to lure food. On the tip of the rod is a luminous organ shaped like a fleshy bulb. The anglerfish waves the lighted organ at its prey and then slowly lays the rod back in a groove on its head, tempting the

A female anglerfish lures prey with its glowing "fishing rod."

prey closer to its gaping mouth. It is not known how the light at the tip of the "fishing rod" is produced, but it has been suggested that luminescent bacteria are responsible.

An amazing feature of one species, the *Ceratias hollbolli*, is the huge difference in size between the female and male. The largest *holl-bolli* male ever found is 4 inches (10 cm) long, while the largest *hollbolli* female is 40 inches (100 cm). In this species, only the female has a fishing rod. She has black rough skin, a pear-shaped body, and a large head and mouth. Because of the female's huge size, she doesn't move much but remains suspended in the water.

By contrast, the *hollbolli* male is tiny, has no lure, and sometimes is without a digestive tube. To survive, the male anglerfish may attach himself to the female, using the teeth on his snout and lower jaw. Once attached, the male begins to shrink in size, and eventually even his eyes disappear. Soon the male becomes fused to the female's body, and their blood vessels join. He is called a **parasite**. Sometimes, two or three males will attach themselves to a female.

The *hollbolli* male is fed through the female's blood vessels, much the same as an unborn baby is fed when it grows inside its mother. This male has no independent existence after that. It does nothing more than **fertilize** any eggs the female produces.

There are more than two hundred species of anglerfish. In some species, the males are free-swimming and do not attach to the females.

CHAPTER 7

Lantern-eye Fish: Built-in Flashlights

Among the many light-producing fish is the lantern-eye fish, or flashlight fish. One species has a bean-shaped light pouch below its eyes and flashes the light on and off. Some species can keep the light on for as long as thirty minutes. The light pouch is actually a series of tubes made up of luminous bacteria. A small eyedropperful (1 ml) of fluid from the pouch contains approximately 10 billion of these continuously glowing bacteria!

Blood vessels in the pouch supply the bacteria with food and oxygen. The partnership of the bacteria and the fish is mutually beneficial. The **host** provides nutrients and protection for the bacteria, which in turn provide a source of light.

Because luminous bacteria cannot be turned on and off, the lantern-eye fish controls

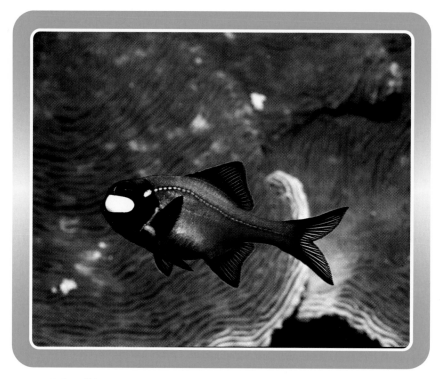

The "lantern" on this photoblepharon appears bright against its dark face.

its light in two ways. One species, the *Photoblepharon*, can draw up a dark lower eyelid, like pulling up a windowshade, over the light pouch. Another species, the *Anomalops*, can rotate the pouch downward into its skin by using special muscles, like retractable headlights in a fancy sports car. By "blinking," it can disappear into its dark world.

Lantern-eye fish range in size from 3 to 12 inches (8 to 30 cm). As some species live near the surface, their light can be seen only at night.

The *Kryptophanaron alfredi* species, found in the Caribbean, live as deep as 600 feet (183 m) below the surface. These fish rise only occasionally to 100 feet (30 m) to feed on **zooplankton**, which are microscopic animals. Lantern-eye fish use the built-in "flashlights" to guide their way in the dark depths, attract prey, confuse predators, and possibly to "talk" with each other. To confuse an enemy, the fish flashes and swims in zigzags. The light strength of the fish's beam equals that of a dim penlight.

Divers easily can catch lantern-eye fish at night. They dive without lights and approach the glowing fish. The divers then stun the fish with a bright light and quickly scoop them into bags.

When luminescent fish are captured and placed in aquarium tanks, they often fail to "light up." **Ichthyologists**, people who study fish, have used such methods as rubbing, electric shocks, and adding **ammonia** to the aquarium's sea water to encourage the fish to shine in captivity. These actions do not hurt the fish but stimulate their natural reactions and allow for scientific study.

CHAPTER 8

Squids:
Shimmering Mist Screens

Of the 400 known species of squids, about 125 are luminescent. Squids exhibit luminescence in three ways: internally, externally, and through the presence of luminescent bacteria.

Internal luminescence is light produced within the tissues of the squid itself by special cell structures called **photophores**. **External luminescence** is light produced in the water through secretions ejected from the squid. **Bacterial luminescence** is light produced on the skin of the squid by colonies of luminescent bacteria living as invited guests in special pockets or areas of the squid.

Each luminous species has its own special distribution of internal light spots, also known as photophores. The light spots can be on the squid's **tentacles**, its eyes, the **mantle** (thin folds

Looking almost like a clown, this reef squid shows off its colorful luminescent skin.

of skin) area, or scattered all over its body. The number of light spots ranges from a dozen to more than a hundred. One Mediterranean species has nearly two hundred lights. Another species shines brilliantly in red, white, and blue. When the animal is examined in daylight, the photophores look like black spots. Squids with internal luminescence are mostly deep-sea dwellers.

In the lower depths of the dark sea, an inky secretion cannot help much, but a cloud of light can. So externally luminescent squids eject the chemicals of the luciferase and luciferin separately into the water. The oxygen in the sea water reacts with the two chemicals to produce a bright, shimmering mist screen. While the enemy is distracted by the glowing bluish green water, the squid escapes to safety. A squid with external luminescence has a double defense plan. In the dark zones, it discharges a cloud of light; in the light zones, a cloud of inky blue.

Seals, sea lions, whales, king penguins, and sea birds often feed on squids. To keep from being eaten, squids confuse their predators, and possibly their sight and smell, by squirting ink clouds.

Squids with bacterial luminescence usually live in shallow waters. The bacteria live under thin folds of skin. The squids have reflector cells and a screen or veil to mask off the light when it is not needed.

Luminescent squids range in size from less than 1 inch (2.5 cm) to 60 feet (18 m). They have eight tentacles plus two arms, all of which have suckers. To propel its way through water, the squid sucks water into a body cavity, then shoots it from an opening behind its head.

Depending on the species, the squid's color can change to red, yellow, blue, purple, or black. It has excellent eyesight. Some species have well-developed eyes similar to ours. Another species has two eyes of different sizes, with tiny lights around the smaller one.

In Japan, around Toyama Bay, firefly squids (*Watasenia*), which are about 5 inches (13 cm) long, are caught during a certain period in late spring and early summer. This torpedo-shaped squid looks like a jeweled pin glittering with blue diamonds as it glides through the water. From about April through June, the small squids rise from the deep for a mating ritual, and hundreds of tons are caught and eaten. The finger-size squids are boiled, fried, broiled, or sun-dried. The uneaten ones are used for fertilizer.

Firefleas: Explosions of Light

Warm shallow waters such as the Caribbean and the sandy shallows off Japan harbor tiny "sea fireflies." These crustaceans, *Cypridina*, are also known as firefleas, probably because of their tiny size of 1/8 inch (0.3 cm), or about the size of a tomato seed. Firefleas swim in clusters and secrete a luminous fluid when disturbed.

To protect themselves from predators, female firefleas spurt cloudlike emergency flashes, like burglar alarms. Male firefleas squirt glowing blue chemical dots in a Morse codelike pattern to attract mates during courtship. After mating, the female firefleas store sperm for long periods and therefore don't need frequent association with males. Males can outnumber females by one hundred to one.

Firefleas secrete luciferase and luciferin from glands on the upper part of the mouth.

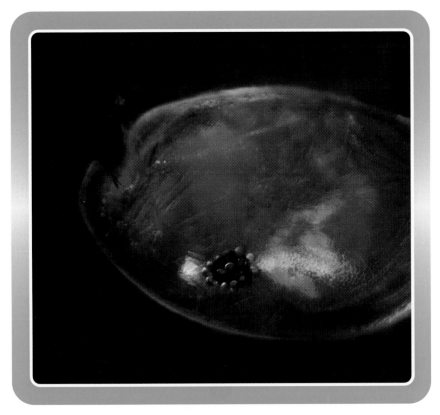

*The fireflea is approximately the same
size as a tomato seed.*

When attacked by small predators, both sexes
explode softball-size clouds of light. The light
startles the predators while the firefleas escape.

There are thirty-nine light-emitting fireflea
species, and each has its own special light pat-

tern. The light dots can be seen through 30 feet (9 m) of water and may last as long as ten to fifteen seconds. Using tiny swimming legs at one end, these animals can swim an amazing sixty body lengths a second. They squirt a trail of light in the water the way a skywriter makes cloud trails in the sky.

During the day, firefleas stay tucked away in the sandy sea bottom. At night, they feed on dead fish. The firefleas are easily caught in fine-meshed **plankton** nets used by marine biologists. Night divers frequently spot their luminescent clouds in the shallow waters above the Caribbean reefs.

The soft-glowing light of *Cypridina* was used by the Japanese during World War II. Large numbers of the luminescent creatures were collected, dried, and powdered. In the jungle, where flashlights or matches were unsafe, a Japanese soldier would moisten and crush some *Cypridina* in the palm of his hand. This created enough dim light to read maps or messages. In dried *Cypridina*, the luminescent chemicals retain their potency for more than twenty-five years after the animal's death. In fact, at Johns Hopkins University in Baltimore, Maryland, they are still using *Cypridina* material collected in the 1920s!

CHAPTER 10
Fireworms: Blazing Water Ballet

The waters off the Bahamas, Bermuda, and Jamaica bring us swarms of "flaming" fireworms. Each month, on the second, third, and fourth days after the full moon, female fireworms gather and swim in circles on the ocean surface, giving off a continuous, glowing yellow-green light. About an hour after sunset, their water ballet show reaches its climax and attracts male fireworms swimming in deep water. Males swim up to the water's surface, giving short flashes of light. The fireworms mingle and form groups of circles. Females release eggs, and males emit sperm. The eggs and sperm combine for fertilization.

Fireworms, or *Odontosyllis*, are 1 to 4 inches (2.5 to 10 cm) in length. The female is continuously luminous. Males have tiny headlights that flash. To find one another, they must wait for

*This fireworm appears to be dressed in a
tutu ready for its water ballet.*

total darkness, which is the interval between
sunset and moonrise. During the summer
months, fireworms spawn several nights in a
row.

These clusters of flashing fireworms are
thought to be the lights seen by Christopher
Columbus in 1492 just before landing in the
New World. The fireworms' instinctive life
cycles seem to be triggered by natural daylight
and the passage of time.

CHAPTER 11
Dinoflagellates: Sparkling Jewels

The ocean currents hold countless tiny swimming and floating creatures called dinoflagellates. These creatures appear greenish white. The largest dinoflagellate is about 1 millimeter, which is the size of the period at the end of this sentence. Each one-celled organism is propelled by two threadlike tails, known as **flagella**.

A huge mass of dinoflagellates floating together is known as a "bloom." During daylight they look like red patches of water. At night, when they are disturbed by passing boats, fish, or human swimmers, they sparkle like jewels. Even a bubble pushing on them causes their tiny membranes to change shape and emit light.

Dinoflagellates exist mostly in salt water. They come in many shapes, and some species are covered with protective plates. They get their

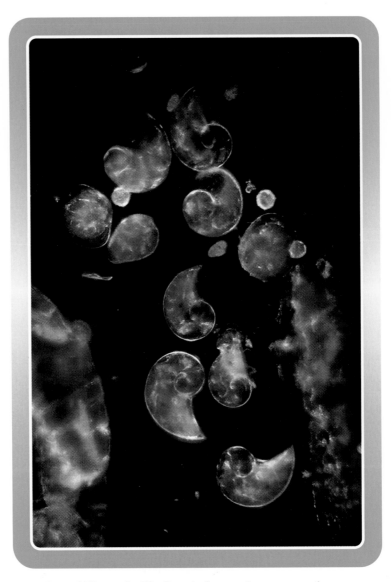

*Sparkling shell-shaped marine organisms
bounce against each other in
Puerto Rico's Phosphorescent Bay*

food from dissolved scraps of leftovers from larger fish.

One well-known area to see dinoflagellates in action is in Puerto Rico's Phosphorescent Bay, on the island's southwest coast. The 60-acre (.24 sq. km) lagoon is attached through a narrow link to the Caribbean's gentle tides and vitamin-rich water.

Every night of the year in the bay, millions of microorganisms glow in the dark. The best displays occur on moonless nights. In this bay, dipping your hand in the water causes it to glow like a ghost. Your wet fingers drip sparks. Boats leave an illuminated trail in the dark waters. Raindrops hitting the water's surface cause the water to shimmer with light.

The *Pyrodinium* dinoflagellates create the beautiful bioluminescent bays in Puerto Rico and Jamaica. Dinoflagellates are animal-like and plantlike. They are animal-like because they can move from place to place. And they are plantlike because they contain chlorophyll, which is a green life-giving pigment.

No one knows why these specklike organisms have built-in lights. No one knows why they glow when the water is agitated or why the light they produce has little or no heat. These answers are buried in each cell's body.

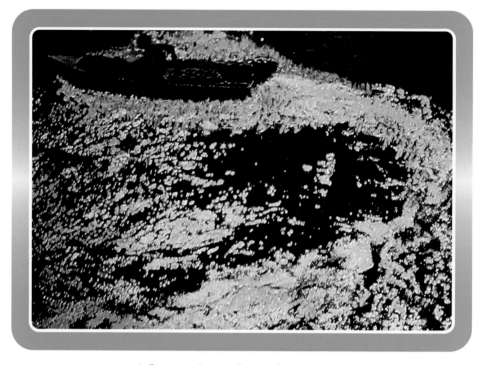

A boater's outboard motor in
Phosphorescent Bay churns glimmering
dinoflagellates on the dark lagoon.

Most Caribbean **phosphorescent** bays have been destroyed by buildings on the coastland. Commercial development has changed the water's quality, quantity, and nutrient supply. According to a 1987 U.S. Navy report, only about fourteen phosphorescent bays remain in the world.

CHAPTER 12

Luciferase and Luciferin Aid Medical Research

Scientists are finding ways to use biolumi-nescence in our everyday life. Medical research has benefited greatly by using the special chemicals of luciferase and luciferin. These chemicals from the firefly help detect diseases in humans.

Bacteria consists of a high amount of the chemical adenosine triphosphate (ATP), an ener-gy-producing compound found in all living things. When a blood or urine sample from an ill person is combined with these chemicals, it will glow brightly if there are bacteria present, showing infection. Cancer cells and tissue dam-age from a heart attack can be detected by inject-ing these special chemicals into the suspected disease area.

Contamination of water, soft drinks, and wines also can be detected by adding a mixture

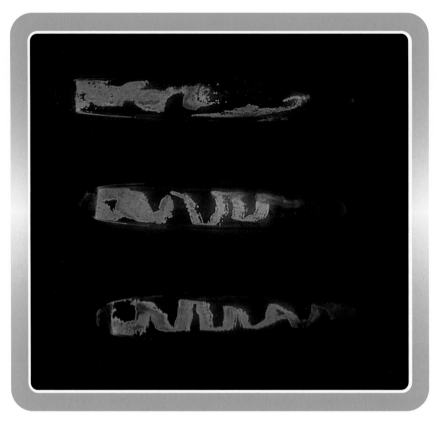

Colonies of luminescent bacteria
are used in research.

of luciferase and luciferin. The presence of bacteria or yeast will cause the liquids to glow.

In the 1960s, researchers discovered how to make luciferin in the laboratory. Scientists are now using bioluminescence to study the way **genes** work within living cells. It wasn't until

1985 that a research team at the University of California in San Diego developed a way to make luciferase. It is grown in special bacteria. This method of copying the luciferase gene is called **cloning**.

The tobacco plant was used by the research team to test cloning of luciferase. The cloned luciferase gene of the firefly was spliced into the tobacco plant cells. When the tobacco plant was "fed" or watered with luciferin, the entire plant, including the roots, glowed a greenish yellow color.

The glowing allows the luciferase gene to be a "reporter." Light is generated when luciferin reacts with the enzyme luciferase and ATP. Reporter genes can be used to identify and control agricultural diseases in plants. The tracing of gene activity in animal and human cells will allow researchers to learn more about **heredity**.

A gift from nature, these living lights are not only fascinating but also lead to discoveries for a better life for us all.

Glossary

Ammonia (eh-MO-ne-eh) — a colorless, sharp smelling gas that dissolves easily in water.

ATP — abbreviation for **adenosine triphosphate** (eh-DEN-o-seen tri-FOS-fate) an energy-producing compound present in all cells.

Bacteria — minute, one-celled organisms.

Bacterial luminescence — minute, single-celled plant organisms, existing as parasites.

Bioluminescence (bi-o-loo-meh-NES-ehns) — the production of light by living things.

Breed — to produce or bring forth young.

Camouflage (KAM-eh-flazh) — to conceal something from the enemy by making it appear to be a section of the natural background.

Clone — all of the descendents of a single individual arising by nonsexual reproduction.

Crustacean (kru-STA-shehn) — any arthropod with a hard, brittle surface or shell.

Cycle — a set of operations that is repeated as a unit.

Enzyme (EN-zime) — a substance produced by living cells that can cause a specific reaction.

External luminescence — light on the outside.

Fertilization — the union of male and female cells, resulting in a new individual; enrichment of the soil to increase crop productivity.

Filament — a minute, threadlike structure or growth form.

Flagella (fleh-JEL-eh) — a slender, whiplike cell that functions as a organ for movement.

Fungi (FUN-ji) — a large group of simple plants that lack chlorophyll, like molds and mushrooms.

Gene — a self-duplicating particle involved in the transmission of genetic information from one generation to the next.

Glowworm — a larva or a wingless beetle that is capable of producing cold light.

Heredity — the capacity of an individual to develop traits present in parents or ancestors.

Host — an organism on or in which another organism lives and for which it provides housing or nourishment or both.

Ichthyology (ik-thee-OL-eh-jee) — the science that deals with fish.

Innards — the internal organs of the body.

Instar — a stage in the life an insect between moltings.

Internal luminescence — light on the inside.

Larva — the wormlike, wingless, immature feeding form which hatches from the egg in insects which undergo complete metamorphosis.

Luciferase (loo-SIF-er-aze) — an enzyme that brings about the oxidation of luciferin.

Luciferin (loo-SIF-er-in) — a substance present in light-producing organs of bioluminescent organisms which, when oxidized, yields light energy.

Luminescence (loo-meh-NES-ehns) — the emission of light without heat.

Luminous — reflecting, emitting, or producing light.

Mandible (MAN-deh-behl) — in arthropods, one or a pair of mouth parts for cutting, crushing, or grinding.

Mantle — thin folds of skin covering the body.

Mucus (MYOO-kehs) — a slimy substance that serves to moisten and lubricate.

Oxygen (OK-si-jen) — an odorless, tasteless, colorless gas essential for life in animals, used in breathing.

Parasite — an organism that lives in or on another organism from which it gets its nourishment.

Phosphorescence (fos-feh-RES-ehns) — the emission of light without heat.

Photophore (FO-teh-for) — a light-producing structure; a luminescent organ.

Plankton (PANGK-tehn) — aquatic organisms of fresh, brackish, or sea water that float passively or exhibit limited locomotor activity.

Predator — one who lives by killing and eating other animals.

Prey — an animal that is seized by another and eaten.

Species (SPEE-seez) — a group of individuals of common ancestry that closely resemble each other structurally and physiologically and in nature, interbreed, producing fertile offspring.

Tentacle — any of various slender, elongated, flexible, unsegmented processes located near the mouth or head of an animal functioning as sensory, food-getting, defensive, or attachment structures.

Zooplankton — microscopic animals that move passively in aquatic ecosystems.

For Further Reading

Harris, Louise Dyer and Norman Dyer Harris. *FLASH—The Life Story of the Firefly*. Boston: Little, Brown, 1966.

Johnson, Sylvia A. *Fireflies*. Minneapolis: Lerner Publications, 1986.

Kohn, Bernice. *Fireflies*. Englewood Cliffs, NJ: Prentice-Hall, 1966.

Poole, Lynn and Gray Poole. *Fireflies in Nature and the Laboratory*. New York: Thomas Y. Crowell, 1965.

Silverstein, Alvin and Virginia Silverstein. *Living Lights — The Mystery of Bioluminescence*. San Carlos, Calif.: Golden Gate Junior Books, 1970.

Silverstein, Alvin and Virginia Silverstein. *Nature's Living Lights*. Boston: Little, Brown, 1988.

Index

About the Author

Judith Janda Presnall grew up in Milwaukee, Wisconsin. She has a bachelor's degree in education from the University of Wisconsin in Whitewater. After twenty years of high school teaching, Judy is now devoting time to writing for children. She and her husband Lance live in Los Angeles, California. They have a daughter, Kaye, and a son, Kory. The California Writers Club has awarded Judy a juvenile nonfiction award.